Real Science -4- Kias

PHYSICS

Connects to Language

DR. R.W. KELLER

Cover design: Rebecca Keller
Opening page: Rebecca Keller
Illustrations: Rebecca Keller

Physics Level I Connects To Language

ISBN # 978-0-9765097-7-6

Published by Gravitas Publications, Inc.
P.O. Box 4790
Albuquerque, NM
87196-4790
www.gravitaspublications.com

Printed in United States

GRAVITAS
PUBLICATIONS INC

Contents

Introduction

1.1 The language of science

Have you ever noticed that scientists use all kinds of fancy words like *nucleosynthesis* [nu'cle-o-syn-the'sis] or *photoelectric photometry* [pho'to-e-lec'tric pho-tom'e-try]. Many of the words that scientists use are long and difficult to pronounce. However, these words have been carefully selected by scientists who have put the field of science into verbal language. Each scientific word means a particular thing. There is a *language* to science.

1.2 Latin and Greek roots

If someone had to memorize all of the words that scientists use it would be a difficult task. However, most of the words that are used in science come to us from two languages: Latin and Greek: Many of the words you encounter will have Latin or Greek *word roots*. A word root is that part of a word that is derived from another word. For example, the word "biology" comes from

the Greek word *bios*, which means "life" and *logy* which means "study of," so biology means the *study of life*. The word tree illustrates several different words and their Latin or Greek word roots. You can see that the words on the branches are the word roots and those on the leaves come from these roots. In fact, many of the languages that people speak have Latin or Greek word roots. English, Spanish, Portuguese, German and even Romanian all have some words that are similar because some words in each of these languages come from Latin or Greek. For example, the English, Spanish, Portugese, French and Italian words for school all come from the Latin word *schola*.

"school"	language
schola	**Latin**
escuela	Spanish
escola	Portugese
scuola	Italian
school	English
ecole	French

We can learn a lot about languages by learning the Latin and Greek word roots.

1.3 How this book works

This workbook is called a "connection" because this workbook *connects* the discipline of science to the language of science. It will help you to understand the different subjects in science if you know more about the language of science.

In the first section of each chapter you will find the word root for a set of six English words.

For example, your word list may look something like this:

deflate

inflate

flabellum

flavor

conflate

afflate

The word root can be two letters long, or three letters long or even four or five letters long. You are asked to look for the two to five letters (or cluster) that are common in each word. This is the word **root**. For example this list of words has the following three letters in common:

deflate

inflate

flabellum

flavor

conflate

afflate

We see that the common cluster in all of the words are the three letters "f," "l," and "a" and make up the word root f l a.

The exercises in the first section are designed to get you thinking about the words in the list and the common word root. You are asked to *guess* the meaning of the root word and any of the words in the list.

In this section the meaning of the word root is defined.
For example, we found that the word root for our example list was the three letter word root **fla**.

deflate   flavor
inflate   conflate
flabellum   afflate

We find out in this section that the meaning for the word root **fla**, comes from the Latin word *flare*, which means "wind," or "to blow." All of the words in the list have something to do with wind or blowing. The exercises in this section encourage you to try to define the words on the list *before* you look at the definitions. Guessing is good! It gets you thinking, even if your guesses are wrong.

The third section defines all of the words. These defintions are taken from a number of different dictionaries including Webster's Unabridged New Twenthieth Century Dictionary 1972 [1], the American Dictionary of the English Language 1828[2], and A Thesaurus of Word Roots of the English Language[3]. Additional Latin or Greek word roots for prefixes or suffixes are also given.

The meanings of the words in our example list are:

Deflate   1. to collapse by letting out air or gas, 2. to lessen the importance of, as with money (*de-* opposite)

Inflate   1. to blow full of air, to expand, 2. to raise the spirts of, to make proud, 3. to increase, raise beyond normal (*in-* in)

Flabellum   1. a large fan usually carried by the Pope, 2. in zoology or botany, the fan-shapped part or structure

Flavor   an odor, smell, or aroma *carried by the wind*

Conflate   1. to blow together, bring together, collect, 2. to combine, melt, fuse, or join (*con-* together)

Afflate   to blow or breathe upon (*af -* to, toward).

In the next section you are given the opportunity to match the definitions of the words to the word list. For example:

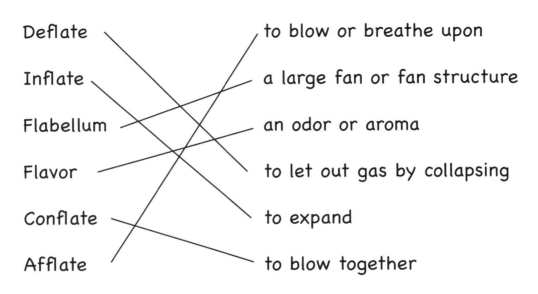

In the next section you "Test Yourself." Now that you have learned the word root and the definitions of the different words in the list you can see how well you remember them with a self-test.

Finally, in the last section of the chapter you will have a chance to make up a story or several sentences using the words you have learned.

For example:

*While carrying a rather **deflated** balloon down the sidewalk, and his hand **conflated** with his mother's, little Johnny was losing his interest with his mother's conversation with Mr. Longs, who awkwardly carrying his **flabellum**, was nevertheless chatting whole-heartedly. The **flavor** of his chocolate sunday was wearing off and his patience was at an end. **Afflating** hard upon his mother, which was his favorite way of gaining her attention, his mother made her farewells and Mr. Longs, still awkwardly carrying his **flabellum**, moved on. Johnny was saved from the boring conversation at last.*

written by Christopher Keller (age 9)

Now that you have learned about how language connects to science, work through this workbook as you learn Level I Chemistry, and most importantly...

have fun!

Chapter 1 : PHYSICS

1.1 Find the root

Look at the following words:

physics

apophysis

metaphysics

physiology

diaphysis

physique

There is a "cluster" or group of letters that is exactly the same in all six words. Can you find the cluster?

Circle the cluster that is the same in each word. Write the four letters that make up the cluster._____ _____ _____ _____.

Now look at the words carefully. Because they all have a common cluster, they all have a meaning with some similarities

Can you guess the meaning of the cluster?

_____

Can you write a definition for any of the words in the list?

_____

_____

_____

_____

1.2 Learn the root

Word Cluster

physics   **phys**iology
apo**phys**is   dia**phys**is
meta**phys**ics   **phys**ique

All of the words above have a common word "root" — **phys**. The word root **phys** comes from the Greek word *physis*, which means "to exist," "to be," or "to grow." All of the words have something to do with "existing," "being" or "growing."

Now, knowing that the cluster **phys** comes from the Greek word *physis,* try to guess the meanings of the words before looking at the definitions in the next section.

physics _____

apophysis _____

metaphysics _____

physiology _____

diaphysis _____

physique _____

1.3 Definitions

physics

the field of science concerned with properties, changes, and interactions of matter

apophysis

a natural process or outgrowth; in geology an outgrowth of igneous rock (*apo*-away from)

metaphysics

a branch of philosophy that deals with the nature of being or reality (*ontology*), the structure and origin of the world (*cosmology*), and the theory of knowledge (*epistomology*) (*meta* - later, behind)

physiology

the branch of biology that deals with how living organisms and their organs and parts function (*logos* - study of)

diaphysis

in anatomy the central part of the bone between the growing ends (*dia*-through)

physique

the structure, form, and appearanace of the body

1.4 Mix and match

Draw lines to connect the words with their meanings.

physics   the structure and form of the body

apophysis   a branch of biology dealing with the
                                     organs and parts of living organisms

metaphysics   the field of science dealing with the
                                     properties and interactions of matter

physiology   an outgrowth

diaphysis   the central part of a bone between the
                                     growing ends

physique   a branch of philosophy dealing with the
                                     nature of reality, the origin of the world
                                     and the theory of knowledge

1.5 Test yourself

Write the meaning next to the words below:

physics _____

apophysis _____

metaphysics _____

physiology _____

diaphysis _____

physique _____

Extra:

Can you guess the meanings of the following words?

physiocrat (*crat*, ruled by)

_____

epiphysics (*epi*, upon)

_____

1.6 Using new words

Write several sentences using each new word you have learned.

_____

_____

_____

_____

_____

_____

_____

_____

_____

_____

_____

_____

_____

_____

Chapter 2 : GRAVITY

2.1 Find the root

Look at the following words:

> gravity
>
> gravitate
>
> aggravate
>
> grave
>
> gravid
>
> gravamen

There is a "cluster" or group of letters that is exactly the same in all six words. Can you find the cluster?

Circle the cluster that is the same in each word. Write the four letters that make up the cluster_____ _____ _____ _____.

Now look at the words carefully. Because they all have a common cluster, they all have a meaning with some similarities.

Can you guess the meaning of the cluster?

_____

Can you write a definition for any of the words in the list?

_____

_____

_____

_____

2.2 Learn the root

Word Cluster

gravity	**grav**itate
ag**grav**ate	**grav**e
gravid	**grav**amen

All of the words above have a common word "root" — **grav**. The word root **grav** comes from the Latin word *gravis*, which means "heavy." All of the words have something to do with "heavy."

Now, knowing that the cluster **grav** comes from the Latin word *gravis,* try to guess the meanings of the words before looking at the definitions in the next section.

gravity _____

gravitate _____

aggravate _____

grave _____

gravid _____

gravamen _____

2.3 Definitions

gravity   in physics, the force that draws two bodies towards each
                   other; weight, heaviness

gravitate   to move according to the force of gravity; to be attracted
                   to or tend to move towards something

aggravate   to make heavy, worse, or severe; to exaggerate (*ad(g)*-to)

grave   (*adj*) requiring serious thought, important, weighty;
                   solemn, sedate; (*n*) a hole in the ground used to bury a
                   body;

gravid   pregnant, with child

gravamen   a complaint or grievance

2.4 Mix and match

Draw lines to connect the words with their meanings.

gravity   being attracted or moving towards
                            something

gravitate   pregnant

aggravate   in physics, the force that brings together
                            two bodies

grave   a complaint or grievance

gravid   to make worse

gravamen   requiring serious thought

2.5 Test yourself

Write the meaning next to the words below:

gravity _____

gravitate _____

aggravate _____

grave _____

gravid _____

gravamen _____

Extra:

Can you guess the meanings of the following words?

gravimeter (*meter*, measure)

_____

gravigrade (*gradus*, to walk)

_____

2.6 Using new words

Write several sentences using each new word you have learned.

Chapter 3 : POTENTIAL

3.1 Find the root

Look at the following words:

> potential
>
> impotent
>
> potentate
>
> despot
>
> omnipotent
>
> potency

There is a "cluster" or group of letters that is exactly the same in all six words. Can you find the cluster?

Circle the cluster that is the same in each word. Write the three letters that make up the cluster_____ _____ _____ .

Now look at the words carefully. Because they all have a common cluster, they all have a meaning with some similarities.

Can you guess the meaning of the cluster?
_____

Can you write a definition for any of the words in the list?
_____

_____

_____

_____

3.2 Learn the root

Word Cluster

potential   im**pot**ent

potentate   des**pot**

omni**pot**ent   **pot**ency

All of the words above have a common word "root" — **pot**. The word root **pot** comes from the Latin word *potis*, which means "able," or "power." All of the words have something to do with being "able," or "powerful."

Now, knowing that the cluster **pot** comes from the Latin word *potis,* try to guess the meanings of the words before looking at the definitions in the next section.

potential _____

impotent _____

potentate _____

despot _____

omnipotent _____

potency _____

3.3 Definitions

potential   something that can, but has not yet, come into being; unrealized, undeveloped

impotent   without power; lacking physical strength; ineffective, helpless (*in(m)*– without)

potentate   a person who possessses great power; a ruler or monarch

despot   master, lord; and absolute ruler; king of unlimited power; tyrant

omnipotent   all powerful; possessing unlimited power (*omni* - all)

potency   the state or quality of being potent or powerful

3.4 Mix and match

Draw lines to connect the words with their meanings.

potential   a tyrant

impotent   something that has not yet come to be

potentate   the state of being powerful

despot   a monarch

omnipotent   without power

potency   all powerful

3.5 Test yourself

Write the meaning next to the words below:

potential _____

impotent   _____

potentate _____

despot   _____

omnipotent _____

potency _____

Extra:

Can you guess the meanings of the following words?

plenipotentiary (*pleni*, full)

_____

potentiometer (*meter*, to measure)

_____

3.6 Using new words

Write several sentences using each new word you have learned.

_____

_____

_____

_____

_____

_____

_____

_____

_____

_____

_____

_____

_____

_____

Chapter 4 : HELIOCENTRIC

4.1 Find the root

Look at the following words:

> heliocentric
>
> parhelion
>
> helium
>
> isohel
>
> Helios
>
> heliotrope

There is a "cluster" or group of letters that is exactly the same in all six words. Can you find the cluster?

Circle the cluster that is the same in each word. Write the three letters that make up the cluster_____ _____ _____ .

Now look at the words carefully. Because they all have a common cluster, they all have a meaning with some similarities.

Can you guess the meaning of the cluster?
_____

Can you write a definition for any of the words in the list?
_____

_____

_____

_____

4.2 Learn the root

Word Cluster

heliocentric   par**hel**ion

helium   iso**hel**

Helios   **hel**iotrope

All of the words above have a common word "root" — **hel**. The word root **hel** comes from the Greek word *helios*, which means "sun." All of the words have something to do with the "sun."

Now, knowing that the cluster **hel** comes from the Greek word *helios*, try to guess the meanings of the words before looking at the definitions in the next section.

heliocentric _____

parhelion _____

helium _____

isohel _____

Helios _____

heliotrope _____

4.3 Definitions

heliocentric   literally "sun centered;" having or taking the sun as the center; calculated or viewed from the center of the sun (*kentron* - center)

parhelion   bright lights near the sun that look like the sun but are not; a fake sun, sometimes connected with one another by a white arc or halo.

helium   the second element on the periodic table, the lightest gaseous element, with an atomic weight of 4.003

isohel   a line drawn on a map that connect two points that receive equal amounts of sunlight

Helios   the Greek sun god

heliotrope   the tendency of certain plants or other organisms to turn or bend under the influence of the sun

4.4 Mix and match

Draw lines to connect the words with their meanings.

heliocentric   the Greek sun god

parhelion   a plant or organism influenced by the sun

helium   sun centered

isohel   the lightest gaseous element

Helios   a fake sun

heliotrope   a line on a map connecting two points
                                      that receive equal sunlight

4.5 Test yourself

Write the meaning next to the words below:

heliocentric _____

parhelion _____

helium _____

isohel _____

Helios _____

heliotrope _____

Extra:

Can you guess the meanings of the following words?

helioscope (*skopein*, view)

_____

heliochrome (*chroma*, color)

_____

4.6 Using new words

Write several sentences using each new word you have learned.

Chapter 5 : NUCLEAR

5.1 Find the root

Look at the following words:

> nuclear
>
> nucleoid
>
> mononucleosis
>
> pronucleus
>
> nucleoplasm
>
> nucleoprotein

There is a "cluster" or group of letters that is exactly the same in all six words. Can you find the cluster?

Circle the cluster that is the same in each word. Write the four letters that make up the cluster_____ _____ _____ _____ .

Now look at the words carefully. Because they all have a common cluster, they all have a meaning with some similarities

Can you guess the meaning of the cluster?
_____

Can you write a definition for any of the words in the list?
_____

_____

_____

_____

5.2 Learn the root

Word Cluster

nuclear   **nucl**eoid

mono**nucl**eosis   pro**nucl**eus

nucleoplasm   **nucl**eoprotien

All of the words above have a common word "root" — **nucl**. The word root **nucl** comes from the Latin word *nux*, which means "nut" or "kernal." All of the words have something to do with the "nut" or "kernal."

Now, knowing that the cluster **nucl** comes from the Latin word *nux,* try to guess the meanings of the words before looking at the definitions in the next section.

nuclear _____

nucleoid _____

mononucleosis _____

pronucleus _____

nucleoplasm _____

nucleoprotein _____

5.3 Definitions

nuclear   a thing or part forming in the center from which something grows; in botany, the kernal or nut or seed; in chemistry and physics, the central part of an atom.

nucleoid   in biology, a region similar to the nucleus found in prokaryotic cells, (*oid* - similar to)

mononucleosios   an infectious disease which creates an excessive number of cells in the blood that contain a single nucleus called leukocytes (*mono* - one)

pronucleus   in zoology the nucleus of either the male or female reproductive cell (gametes) before they combine to create a fertilized egg

nucleoplasm   a plasm (fluid material) inside the nucleus of a cell

nucleoprotein   a protein found inside the nucleus of a cell

5.4 Mix and match

Draw lines to connect the words with their meanings.

nuclear   an infectious disease that causes an
                                 excess number of leukocytes

nucleoid   a protein found in the nucleus of a cell

mononucleosis   the kernal or seed

pronucleus   the fluid material inside the nucleus of a
                                 cell

nucleoplasm   the nucleus of either male or female
                                 gametes

nucleoprotein   a region similar to the nucleus found in
                                 prokaryotic cells

5.5 Test yourself

Write the meaning next to the words below:

nuclear _____

nucleoid _____

mononucleosis _____

pronucleus _____

nucleoplasm _____

nucleoprotein _____

Extra:

Can you guess the meanings of the following words?

nucleonics (-*nics*, dealing with electro*nics*)

_____

nucleoform (*forma*, form)

_____

5.6 Using new words

Write several sentences using each new word you have learned.

_____

_____

_____

_____

_____

_____

_____

_____

_____

_____

_____

_____

_____

_____

Chapter 6 : ELECTRIC

6.1 Find the root

Look at the following words:

electric

electrode

dielectric

electrolysis

pyroelectric

electrophilic

There is a "cluster" or group of letters that is exactly the same in all six words. Can you find the cluster?

Circle the cluster that is the same in each word. Write the six letters that make up the cluster ___ ___ ___ ___ ___ ___ .

Now look at the words carefully. Because they all have a common cluster, they all have a meaning with some similarities.

Can you guess the meaning of the cluster?
_____

Can you write a definition for any of the words in the list?
_____

_____

_____

_____

6.2 Learn the root

Word Cluster

electric	**electr**ode
di**electr**ic	**electr**olysis
pyro**electr**ic	**electr**ophilic

All of the words above have a common word "root" — **electr**. The word root **electr** comes from the Greek word *electrum*, which means "shining, amber." The original meaning is uncertain, but is related to electrons.

Now, knowing that the cluster **electr** comes from the Greek word *electrum*, try to guess the meanings of the words before looking at the definitions in the next section.

electric _____

electrode _____

dielectric _____

electrolysis _____

pyroelectric _____

electrophylic _____

6.3 Definitions

electric
: charged with or conveying electricity; producer or produced by electricity; operated by electricity; (electricity is a form of energy where electrons are moved from one place to another as in a wire)

electrode
: either of the two terminals, either positive or negative, in a battery (*ode-* way, path)

dielectric
: an insulator such as glass or rubber that does not permit the flow of electrons (*dia* - through, across)

electrolysis
: 1. in chemistry, the breakdown of a substance into ions by the action of an electric current passing through 2. the removal of unwanted hair with an electric needle (*lyein* - remove, loosen)

pyroelectric
: the development of electric polarity by a change in temperature (*pyro* - fire)

electrophilic
: a chemical or compound that accepts electrons (*phila*-love)

6.4 Mix and match

Draw lines to connect the words with their meanings.

electric   a substance that will develop electric
                            polarity because of heat

electrode   an insulator

dielectric   electron loving

electrolysis   conveying electricity

pyroelectric   the positive or negative terminal
                            of a battery

electrophilic   the breakdown of a compound into ions
                            by an electric current

6.5 Test yourself

Write the meaning next to the words below:

electric _____

electrode _____

dielectric _____

electrolysis _____

pyroelectric _____

electrophilic _____

Extra:

Can you guess the meanings of the following words?

photoelectric (*photo*, light)

_____

hydroelectric (*hydro*, water)

_____

6.6 Using new words

Write several sentences using each new word you have learned.

_____

_____

_____

_____

_____

_____

_____

_____

_____

_____

_____

_____

_____

_____

Chapter 7 : THERMAL

7.1 Find the root

Look at the following words:

> thermal
> diathermy
> exothermic
> hypothermia
> thermometer
> endothermic

There is a "cluster" or group of letters that is exactly the same in all six words. Can you find the cluster?

Circle the cluster that is the same in each word. Write the five letters that make up the cluster ____ ____ ____ ____ ____.

Now look at the words carefully. Because they all have a common cluster, they all have a meaning with some similarities.

Can you guess the meaning of the cluster?

_____

Can you write a definition for any of the words in the list?

_____

_____

_____

_____

7.2 Learn the root

Word Cluster

thermal   dia**therm**y

exo**therm**ic   hypo**therm**ia

thermometer   endo**therm**ic

All of the words above have a common word "root" — **therm**. The word root **therm** comes from the Greek word *therme*, which means "heat." All of the words have something to do with "heat."

Now, knowing that the cluster **therm** comes from the Greek word *therme,* try to guess the meanings of the words before looking at the definitions in the next section.

thermal _____

diathermy _____

exothermic _____

hypothermia _____

thermometer _____

endothermic _____

7.3 Definitions

thermal   having to do with heat; warm or hot; such as *thermal underwere thermal capacity, thermal unit* etc.

diathermy   in medicine, a procedure where heat is produced in the tissues of the skin by an electric current (*dia*—through, across)

exothermic   in chemistry, a chemical reaction that gives off heat (*exo*—out, outward)

hypothermia   in medicine, a body temperature that is below normal (*hypo*—under)

thermometer   an instrument that measures temperature (*meter*—to measure)

endothermic   in chemistry a reaction that consumes, or takes in heat (*endo* — within, inner)

7.4 Mix and match

Draw lines to connect the words with their meanings.

thermal   a body temperature below normal

diathermy   having to do with heat

exothermic   a chemical reaction that consumes heat

hyopothermia   a chemical reaction that gives off heat

thermometer   in medicine, a procedure that produces
                                     heat in skin tissue

endothermic   an instrument that measures heat

7.5 Test yourself

Write the meaning next to the words below:

thermal _____

diathermy _____

exothermic_____

hypothermia _____

thermometer _____

endothermic _____

Extra:

Can you guess the meanings of the following words?

homoiothermous (*homoio*, same, similar)

_____

poikilothermous (*poikilos*, various)

_____

7.6 Using new words

Write several sentences using each new word you have learned.

_____

_____

_____

_____

_____

_____

_____

_____

_____

_____

_____

_____

_____

Chapter 8 : INDUCTION

8.1 Find the root

Look at the following words:

> induction
>
> deduce
>
> reduce
>
> product
>
> aqueduct
>
> transducer

There is a "cluster" or group of letters that is exactly the same in all six words. Can you find the cluster?

Circle the cluster that is the same in each word. Write the three letters that make up the cluster ___ ___ ___ .

Now look at the words carefully. Because they all have a common cluster, they all have a meaning with some similarities.

Can you guess the meaning of the cluster?

_____

Can you write a definition for any of the words in the list?

_____

_____

_____

8.2 Learn the root

Word Cluster

in**duc**tion   de**duce**

re**duce**   pro**duc**t

aque**duc**t   trans**duc**er

All of the words above have a common word "root" — **duc**. The word root **duc** comes from the Latin word *ducere*, which means "to lead" or "to pull." All of the words have something to do with "leading" or "pulling."

Now, knowing that the cluster **duc** comes from the Latin word *ducere,* try to guess the meanings of the words before looking at the definitions in the next section.

induction _____

deduce_____

reduce _____

product _____

aqueduct _____

transducer _____

8.3 Definitions

induction   leading or bringing into; in physics the act of bringing
                an electric or magnetic effect by some field or force (*in*
                —towards, within)

deduce   to draw conclusions from premises; infer; gather (*de*
                —from)

reduce   to diminish, decrease, shorten; pull back (*re* — back again)

product   something that has been "led forward;" exhibited; (*pro*
                —forward, before)

aqueduct   a channel leading water to a new place (*aqua* — water)

transducer   in physics a device that leads power from one system
                through something to another system (*trans* — through)

8.4 Mix and match

Draw lines to connect the words with their meanings.

induction   a device that leads power from one place
                                 to another

deduce   a channel that leads water from one
                                 place to another

reduce   leading or bringing into

product   to draw conclusions from a set of
                                 premises

aqueduct   something that has been led forward or
                                 exhibited

transducer   diminish, decrease

8.5 Test yourself

Write the meaning next to the words below:

induction   _____

deduce   _____

reduce _____

product   _____

aqueduct   _____

transducer   _____

Extra:

Can you guess the meanings of the following words?

educate (*ex(d)*, out)

_____

viaduct (*via*, road)

_____

8.6 Using new words

Write several sentences using each new word you have learned.

_____

_____

_____

_____

_____

_____

_____

_____

_____

_____

_____

_____

_____

_____

Chapter 9 : DECIBEL

9.1 Find the root

Look at the following words:

> decibel
>
> decanal
>
> duodecimal
>
> hexadecane
>
> decimeter
>
> December

There is a "cluster" or group of letters that is exactly the same in all six words. Can you find the cluster?

Circle the cluster that is the same in each word. Write the three letters that make up the cluster ____ ____ ____ .

Now look at the words carefully. Because they all have a common cluster, they all have a meaning with some similarities.

Can you guess the meaning of the cluster?

_____

Can you write a definition for any of the words in the list?

_____

_____

_____

_____

9.2 Learn the root

Word Cluster

decibel   **dec**anal

duo**dec**imal   hexa**dec**ane

decimeter   **Dec**ember

All of the words above have a common word "root" — **dec**. The word root **dec** comes from the Latin word *decem*, which means "tenth" or "ten." All of the words have something to do with "ten."

Now, knowing that the cluster **dec** comes from the Latin word *decem,* try to guess the meanings of the words before looking at the definitions in the next section.

decibel _____

decanal _____

duodecimal _____

hexadecane_____

decimeter _____

December _____

9.3 Definitions

decibel   a unit for measuring the volume of sound; ten decibels is equal to one bel.

decanal   pertaining to a dean; [*dean* was originally the term for a person in charge of ten monks, later soldiers.]

duodecimal   a system of numbering with twelve as its base [*duo* - two; *duo + dec*, two plus ten equals 12]

hexadecane   a hydrocarbon with sixteen atoms of carbon (*hexa* — six; *hexa + dec*, six plus ten equals sixteen)

decimeter   a unit of measure equal to one tenth of a meter (*meter* — to measure)

December   the tenth month of the early Roman calendar; today the 12th month in the year

9.4 Mix and match

Draw lines to connect the words with their meanings.

decibel   a dean

decanal   the tenth month of the Roman calendar

duodecimal   a hydrocarbon with sixteen carbon atoms

hexadecane   a numbering system with base twleve

decimeter   1/10 of a bel

December   one tenth of a meter

9.5 Test yourself

Write the meaning next to the words below:

decibel _____

decanal _____

duodecimal _____

hexadecane _____

decimeter _____

December _____

Extra:

Can you guess the meanings of the following words?

dime (*hidden root*)

_____

dozen (*hidden root*)

_____

9.6 Using new words

Write several sentences using each new word you have learned.

_____

_____

_____

_____

_____

_____

_____

_____

_____

_____

_____

_____

Chapter 10 : PETROLEUM

10.1 Find the root

Look at the following words:

> petroleum
>
> petrify
>
> petrology
>
> petroglyph
>
> petrous
>
> petrogenic

There is a "cluster" or group of letters that is exactly the same in all six words. Can you find the cluster?

Circle the cluster that is the same in each word. Write the four letters that make up the cluster ___ ___ ___ ___.

Now look at the words carefully. Because they all have a common cluster, they all have a meaning with some similarities.

Can you guess the meaning of the cluster?
_____

Can you write a definition for any of the words in the list?

_____

_____

_____

_____

10.2 Learn the root

Word Cluster

petroleum   **pet**rify

petrology   **petr**oglyph

petrous   **petr**ogenic

All of the words above have a common word "root" — **petr**. The word root **petr** comes from the Greek word *petra*, which means "rock." All of the words have something to do with "rock."

Now, knowing that the cluster **petr** comes from the Greek word *petra,* try to guess the meanings of the words before looking at the definitions in the next section.

petroleum _____

petrify _____

petrology _____

petroglyph _____

petrous _____

petrogenic _____

10.3 Definitions

petroleum   a flammable oily liquid that comes out of the ground composed largely of hydrocarbons and refined for fuel (*oleum* — oil).

petrify   to convert organic matter, such as bone, tissue, or wood, into stone

petrology   the study of the composition, structure, and origin of rocks (*logy* — to study)

petroglyph   any marking, inscription, or drawing etched into the face of a rock or cliff (*glyph* — carving)

petrous   of or like a rock; hard, stony

petrogenic   something that produces rocks (*gen* — producing)

10.4 Mix and match

Draw lines to connect the words with their meanings.

petroleum   like a rock

petrify   the study of rocks

petrology   an oily liquid found in rocks that is used
                                   for fuel

petroglyph   to make into stone

petrous   rock producing

petrogenic   an etching or drawing found in rocks

10.5 Test yourself

Write the meaning next to the words below:

petroleum _____

petrify _____

petrology _____

petroglyph _____

petrous _____

petrogenic _____

Extra:

Can you guess the meanings of the following words?

Peter (*hidden root*)

_____

lamprey (*hidden root*)

_____

10.6 Using new words

Write several sentences using each new word you have learned.

_____

_____

_____

_____

_____

_____

_____

_____

_____

_____

_____

_____